Dr. Peter Ubah Okeke

Distribution of Intestinal Parasitic Infections among the Residence of Porto Novo Municipality of Cape Verde

GRIN Publishing

Bibliographic information published by the German National Library:

The German National Library lists this publication in the National Bibliography; detailed bibliographic data are available on the Internet at http://dnb.dnb.de .

Imprint:

Copyright © 2012 GRIN Verlag, Open Publishing GmbH
Print and binding: Books on Demand GmbH, Norderstedt Germany
ISBN: 978-3-656-19976-2

This book at GRIN:

http://www.grin.com/en/e-book/194273/distribution-of-intestinal-parasitic-infections-among-the-residence-of

GRIN - Your knowledge has value

Since its foundation in 1998, GRIN has specialized in publishing academic texts by students, college teachers and other academics as e-book and printed book. The website www.grin.com is an ideal platform for presenting term papers, final papers, scientific essays, dissertations and specialist books.

Visit us on the internet:

http://www.grin.com/

http://www.facebook.com/grincom

http://www.twitter.com/grin_com

DISTRIBUTION OF INTESTINAL PARASITIC INFECTIONS AMONG THE RESIDENCE OF PORTO NOVO MUNICIPALITY OF CAPE VERDE.

By

Dr. Peter Ubah Okeke

ACKNOWLEDGMENT

I use this forum to thank Arlindo Maocha, who helped me effectively in the collection of stool samples. Dr. Linda Collazo, of the Atlantic International University, Academic department, for all her assistance throughout my stay at AIU programs.

TABLE OF CONTENTS

ABSTRACT

BACKGROUND: Intestinal parasitic diseases are among the most common diseases globally and for this reason, it is very important to study its occurrence and its degree of risk in contamination. No previous record is available in the municipality of Porto Novo, hence the necessity of this study.

AIMS: To obtain the occurrence of common intestinal parasites among the inhabitants of Porto Novo municipality.

Stool testing performed on 391 subjects between March to June, 2011 age ranging from 11 months to 82 years old, 141 (36.06%) were infected with one or more of the intestinal parasites. Among protozoa, Entamoeba coli (22.51%), Entamoeba Histolytica/Dispar (7.67%) and Giardia Lamblia (5.90%) were the most isolated intestinal parasites and among the helminths, Hymenolepis nana (2.30%), Ancylostoma Duodenale (1.28%) and Trichuris Trichuria (0.51%) were isolated. Age distribution did not show a definite pattern of infectivity rather females (24.60%) were mainly infected than males (11.50%).

Hence, concluded that intestinal parasites pose a serious public health problem in the municipality of Porto Novo and that its degree of contamination is still high, therefore, treatment measures, National deworming programs at schools and sanitary improvement strategies is advocated for the population of the municipality to reduce and or eradicate this sporadic problem.

KEY WORDS: Distribution, Intestinal Parasites, Municipality, Porto Novo, Cape Verde.

INTRODUCTION

A parasite is an organism that is entirely dependent on another organism, referred to as its host for all or part of its life cycle and metabolic requirements. Parasitism is therefore a relationship in which a parasite benefits and the host provides the benefit. The degree of dependence of a parasite on its host varies.

An obligatory parasite is one that must always live in contact with its host. The term free- living describes the non parasitic stages of existence which are lived independently of a host for example, hookworms, have active free living stages in the soil.

Definitive host is the host in which sexual reproduction takes place or in which the most highly developed form of a parasite occurs. When the most mature form is not obvious, the definitive host is the mammalian host. However, the intermediate host refers to the host which alternates with the definitive host and in which the larval or asexual stages of a parasite are found. Some parasites require two intermediate hosts in which to complete their life cycle.

The reservoir host is an animal host serving as a source from which other animals can become infected. Epidemiological speaking, reservoir hosts are important in the control of parasitic diseases. The can maintain a nucleus of infection in an area. The term zoonosis is used to describe an animal infection that is naturally transmissible to humans either directly or indirectly via a vector (insect).

Gastrointestinal parasites are frequently transmitted via food and contaminated drinking water but may also be spread from person to person through faecal- oral contact. It is believed that over 70 species of protozoan and helminthic parasites can infect human through food and water contamination (Pozio 2003).

A third of the world´s population most of them children may be infected with intestinal worms, principal among them are Ascaris Lumbricoides, Hookworms, and Trichuris Trichuria which cause a variety of conditions including malnutrition, iron deficiency anaemia, malabsorption syndrome, intestinal obstruction and mental and physical growth retardation (Allen & Maizel 1996).

The protozoan Entamoeba Histolytica cause of amoebiasis and life threatening liver abscess is a leading cause of death due to parasites second only to malaria (Stanley 2003). Although many intestinal parasites, particularly the geohelminths have virtually disappeared from industrialized countries, screening for parasitic infections remains a public health priority in the United States of America health care settings serving refugees and immigrants(Walker & Jaranson 1999), most of whom may have emigrated from countries where intestinal parasitic diseases are endemic(Larson 2003).

Helminthes and protozoan infections differ importantly both in host immune response and in epidemiology.Helminths are macroparasites that reproduce sexually within the definitive host, where the can persist for many years, whereas protozoa are microparasites that are capable of direct reproduction within the host, often at high rates and cause relatively short lived infection (Maizel 1993).

Helminth infections are well known to elicit a type-2 (th-2) non inflammatory T-cell response, a hallmark of which is IgE elevation with eosinophilia and which may be strong enough to exert biasing effects on concomitant infections (Pit DS 2001), as well as host response to Th-1 mediated chronic infections, including intracellular parasites. By contrast, many protozoa like bacteria act through manipulation of Th-1 pathways.

In developing countries, particularly those with tropical climates and at low altitudes, such infections remain a serious medical and public health concern. They are more prevalent among the poor, who are negatively affected by low socio- economic conditions, poor personal and environmental hygiene, overcrowding and limited access to clean water (Mengistu et al 2007).

In Cape Verde and elsewhere, food vendors and restaurants are noted for selling foods and drinks at reduced rate, so providing more affordable means for people to obtain the balanced diets outside the home. An estimated 2.5 billion people patronize food vendors worldwide (Nyarango et al 2003). Although street food has become an indispensable part of both urban and rural diets, however, in developing countries, some public health risks are associated with the consumption of street food. While it is surely expected that street food meets the nutritional needs of consumers, it is also vital to ensure its safety from contaminants and microorganisms (Chakravarty 2001).

Illness due to contaminated food has also been reported by the World Health Organization (WHO) as the most widespread health problem in the contemporary world and an important cause of reduced economic productivity (Kaferstein 2003). Food borne illness can therefore be considered a major international health problem and an important cause of reduced economic growth (FAO/WHO 1983).

These parasitic diseases whether food borne, water borne, vector borne, soil transmitted or those that result from some poor sanitary or social habits provide some of the many public health problems in the tropics (Woodrouff,1965 & Odutan, 1974).

The disease process which emanates may be the consequences of the reactions of human host to the parasites invading the host´s tissue, causing destruction and damage to the tissues, or the result of the parasites depriving the human host of some essential nutrients. Parasitic diseases create morbidity and sometimes mortality. Estimates of these parasitic diseases in the Population of Porto Novo thus become a matter of necessity for the surveillance of public health, proper health care delivery and people´s welfare.

MATERIALS AND METHODS

STUDY AREA

This study was undertaken in the medical laboratory section of Central Hospital Porto Novo.

SUBJECTS

They subjects were taken from the municipality of Porto Novo. They are apparently healthy individuals from all around the municipality and those visiting the Hospital.

COLLECTION OF SAMPLE

Each patient were given a clean, dry, well labeled specimen container and asked to collect and submit two samples of stool on different days of the week and or during the period of the research, which started 1-March 2011 to 30-June, 2011.

LABORATORY TESTING OF STOOL

In the laboratory, each specimen was first examined macroscopically for its consistency, colour and presence of blood, mucus, adult worm and proglottids of taenia spp.With the aid of an applicator stick, normal saline and Lugol's iodine preparation were made for direct examination or wet mount examination. The sodium chloride floatation method or the technique of Willis and the modified Ziel Neelsen staining method were used according to Cheesbrough and also WHO standard testing for Helminth ovas were used. The antigen test for the differentiation of Entamoeba Histolytica and Entamoeba Dispar were not possible.

RESULTS

Out of a total of 391 subjects examined, 141 (36.06%) were infected with one or more intestinal parasites. Age distribution did not show a definite pattern but infection rate were higher among the age range of 32 to 47 years old with 12.53% prevalence, although this is not statistically significant ($p<0.05$). Despite that the subjects were not diarrheal patients and do not manifest major symptoms of intestinal parasitic disease, infection rate was relatively high.

Entamoeba coli was the most occurred parasite(22.51%) while Giardia lamblia (5.90%), Entamoeba Histolytica/Dispar (7.67%), Iodamoeba butschlli(4.09%), Balantidum coli and Endolimax nana were 0.30% and 0.51% respectively. The helminths isolated were Hymenolepis nana (2.30%), Ancylostoma duodenale (1.28%) and Trichuris Trichuria (0.51%).

Table-1: Occurrence of parasites in 391 stool samples from the residence of Porto Novo municipality of Cape Verde, March to June, 2011.

Parasites	N°	Percentage (%)
PROTOZOA		
Entamoeba coli*	88	22.51
Entamoeba Histolytica/Dispar	30	7.67
Giardia lamblia	23	5.90
Iodamoeba butschlli*	16	4.09
Endolimax nana*	2	0.51
Balantidum coli	1	0.30
HELMINTHS		
Hymenolepis nana	9	2.30
Ancylostoma duodenale	5	1.28
Trichuris Trichuria	2	0.51

*Commensal, non-pathogenic.

Table-2. Age-related occurrence of intestinal parasitic infection of a total of 391 stool specimen obtained during the Porto Novo municipality study, March to June, 2011

Age in years	N° infected	Percentage (%)
0-15	46	11.80
16-31	33	8.44
32-47	49	12.53
>48	13	3.32

Table- 3: Sex related distribution of intestinal parasitic infections of the 391 faecal samples from the residence of Porto Novo municipality study, March to June, 2011.

Gender	N° Infected	Percentage (%)
Male	45	11.50
Female	96	24.60

Table-4: Occurrence of mono and Polyparasitism of 391 faecal specimen of the inhabitants of municipality of Porto Novo between March to June, 2011.

Stool samples examined (n: 391)	Number	Percentage (%)
Negative	250	63.94
Monoparasitism	107	27.37
Polyparasitism	34	8.70

DISCUSSION AND CONCLUSION

The results of 36.06% showed in this study confirmed that intestinal parasitic infection is still a major public health problem in Porto Novo municipality. The commonest parasites isolated was Entamoeba coli (22.51%) and Entamoeba Histolytica /Dispar 7.67%, Giardia lamblia infection was 5.90% and this occurred mainly in children aged 11 months to 8 years old. The helminths accounted for 4.0% which was not statistically significant (p<0.05), among the worms, Hymenolepis Nana was 2.30%, Ancylostoma Duodenale registered 1.28% and Trichuris Trichuria was 0.51%.

The Porto Novo municipality experiment showed that parasitic infection due to helminths is not common. Although, World Health Organization stated that intestinal protozoan parasites are age dependent and greater severity of the infection is found in children. This could be attributed to the different host responses and nutritional status. The Porto Novo experience showed that age related occurrence of parasitic infection is not statistically significant (p<0.05). However, Giardia lamblia infection was in children than in adults possibly due to daycare schools allow the children to be exposed to the infecting forms and this was noted in age 11 months to 8 years old. The hygienic conditions and the low immune responses of these children to the parasites could also be a determining factor to them been infected. According to the Porto Novo testing on parasites, females (24.60%) were more infected than males (11.50%) and this is directly in agreement to the work of Atu et al (2006) that reported higher occurrence of intestinal parasites in females than males. Also in this study, 27.37% of those infected were Monoparasitism whereas 8.70% were of Polyparasitism.

The high occurrence of Entamoeba coli (22.51%), a Commensal parasite is indicative of the population's precarious sanitary conditions and of elevated environmental contamination, highlighting the need for education focused on hygienic measures, along with investments on sanitation.

The water supply is really an important risk factor for amoebiasis and giardiasis and several large outbreaks have resulted from the contamination of municipal water supplies with human waste as was reported by Wilson et al (1998).

In view of the occurrence of intestinal parasitic infections of 36.06% recorded in this work, preventive measures becomes imperative to prevent dissemination of infection and or to reduce opportunities for the exposure, by increasing the level of knowledge about personal and community health and hygienic sanitary control of water and waste disposal and reduction of the source of infection by therapeutic means.

In conclusion, treatment measures should be taken for those already infected. Sanitary improvements such as safe, efficient and dynamic hygienic management of drinking water and disposal of excreta should be undertaken. Regular hand washing with soap and water should become a habit and taught at the basic schools and other sectors and the necessary materials for this should be provided. Hygienic food preparation, handling, storage and health education by the public health officers to encourage individuals to adopt behavioral change is advocated. Regular inspection of all food processing and or selling areas by the public health officers of the local council should be encouraged and or practiced.

The program of intermittent deworming of school children by the ministry of health of Cape Verde or its delegated health officers should be re-enforced and this could be effected twice in a year. At all it seems that a multisectoral control approach is needed in the Porto Novo municipality of Cape Verde to control, screen and totally eradicate intestinal parasitic infection in our region.

REFERENCE

Adekunle L (2002): Intestinal parasites and nutritional status of Nigerian children. Afr. J. Biomed. Res; 5:115-119.

Agi PL (1995): Pattern of infection of intestinal parasites in Sagbama community of the Niger Delta, Nigeria. West Afr J Med 14:39-42.

Allen JE et al: Immunology of human Helminth infection. Int. Arch. Allergy I mmunol. 1996; 109:3-10.

Atu B.O et al (2006): Prevalence of intestinal parasites in Etulo, Benue state Nigeria. Nig. J. of Parasitology, 7:1-16

Chan MS et al: The evaluation of potential global morbidity due to intestinal nematode infection. Parasitol.1994; 109:373-87.

De Silva NR et al: Soil transmitted Helminth infection. Updating the global picture. Trends Parasitol.2003; 19:547-51.

FAO/WHO: The role of food safety in health and development. Report of the joint FAO/WHO expert committee on food safety. WHO 1983.

Henry J B: Clinical diagnosis and management by laboratory method, 17th edn. Philadelphia: WB Saunders; 1995.

Kaferstein FK: Actions to reverse the upward curve of food borne illness. Food control 2003; 14:101-9.

Kim BJ et al (2003): The intestinal parasite infection status of the inhabitants in the Roxas city, The Philippines. Korean J parasitol.41:113-115.

Maizel RM et al: Immunological modulation and evasion by Helminth parasites in human population. Nature, 1993; 365:797-805.

Manual of basic techniques for a health laboratory 2nd edn. WHO 2003.

Mengistu A et al: Prevalence of intestinal parasitic infections among urban dwellers in south west Ethiopia. Ethiopia J health Dev. 2007; 21:12-7.

Monica Cheesbrough: District laboratory practice in tropical countries part 1, 1999:183-193.

Nyarango RM et al: The risk of pathogenic parasite infection in Kissii, Kenya.BMC Public Health.2003; 8:237.

Obeng AS et al: Parasitic pathogen microbes associated with fresh vegetables consumed in Accra. Ghana J Allied health Sci. 2007; 2:11-5.

Odutan SO: Housing and health of Nigerian children.Trop. Geogr. Soc. Med. Parasitol. 1974;25:402-409.

Pit DS et al: Parasite specific antibody and cellular immune responses in human infected with Nector Americanus and oesophagostomum Bifurcum. Parasitol.Res. 2001; 87: 722-729.

Pozio E: Food borne and water borne parasite. Acta microbial pol. 2003;52: 83-96

Stanley SL: Amoebiasis. Lancet 2003; 361:1025-34.

Stephenson LS et al: Malnutrition and parasitic Helminth infections. Parasitl. 2002; 121:23-38.

Walker PF et al: Refugee and immigrant health care. Med Clin North American 1983:1103-1120.

Woodrouff AW: Infection with animal Helminthes. Br. Med. J. 1965 pp.1001.

World Health Organization (2005): Deworming for the population and development. Report of the third global meeting of the partners for parasite control see www.who.int.